FIELD
PHILOSOPHY

Anthony Pace

Ultimate Identity
Explanatory Of
The Cosmos

To order additional copies of this book, contact:
Xlibris
844-714-8691
www.Xlibris.com
Orders@Xlibris.com

ISBN: 979-8-3694-1952-6 (sc)
ISBN: 979-8-3694-1951-9 (e)

Library of Congress Control Number: 2024907200

Print information available on the last page

Rev. date: 04/27/2024

ARMADILLO: NONFICTION

APPROXIMATE DIMENSIONS:

LENGTH	10'
WIDTH	2'6"
HEIGHT	3'

Summer of 1969, 5:00 am. Wallingford, CT. Nearby a red clay plateau. Sighting of a 7 plate armadillo; 'Natuarl Philosophy' study of one release 7 plate armadillo; 'Moral Philosophy'. Ethical, study of the observable realm of changing cloud molecular particulars known as goddess of Lyra or Feynmans sprite on the stair case; 'Metaphysical Philosophy' Abstruse, as of the 45° plateau in likeness of Lyra 45° angle;

Anthony Pace

Ultimate Identity Explanatory Of The Cosmos

Anthony Pace

Ultimate Identity Explanatory Of The Cosmos

Archives October 2010

From the top young musicians.org Harpiest wrote the song Doll Monster from the poster

At 4 years of age every morning use to rub an Arabian spirit genie. Next to a window overlooking the clay plateau plane where the Goddess was sighted.

Myth of a fairy return as an armadillo

FAIRYLAND March

REALM,
OF THE HEAVENS,
FAIRY

30 cm

6

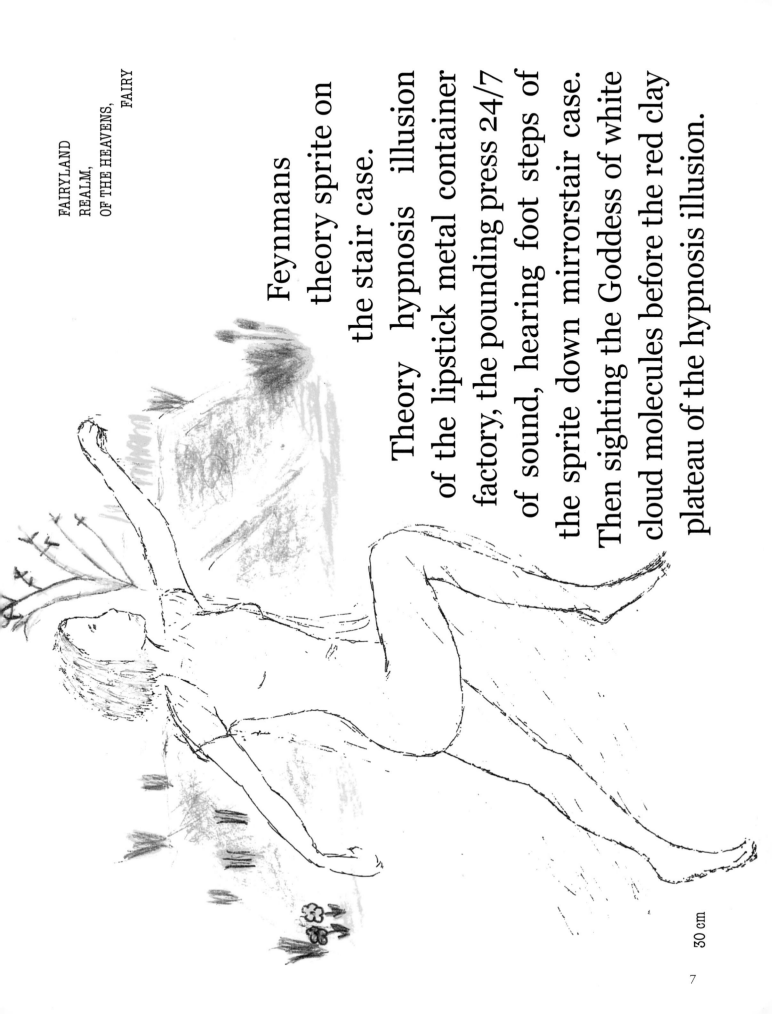

Feynmans
theory sprite on
the stair case.

Theory hypnosis illusion
of the lipstick metal container
factory, the pounding press 24/7
of sound, hearing foot steps of
the sprite down mirrorstair case.
Then sighting the Goddess of white
cloud molecules before the red clay
plateau of the hypnosis illusion.

30 cm

Goddess of Lyra

REALM, OF THE HEAVENS, FAIRY

Field Philosophy

Mapping of an Ice Planet like Mars at Lyra 1976, Philosophy- Natural, Moral, Metaphysical:

Famous Philosophy

If you sighted an mirage in the desert; What would you do with it?

30 cm

On the Egg Farm Nov. 14, 2004 Sunday, 9:00 P.M.. Sighted a meteorite that turned into a tortuous of orange gas. That looked very beautify and came towards Earth and turned out into space and burnt out.

At that time NASA was asking for people who sighted meteorites, so I sent the drawing to Nasa. At that time there was a contest for a spacecraft. This tortuous meteorite would be perfect for a spacecraft of aerodynamics in outerspace.

July 22, 2005

Summer on the Egg Farm sighted a Thunder Bolt that looked like a mirror of mercury Jello. Came from the cloud to hitting the Earth of field. Stayed from the cloud to the field for about a minute. When the Thunder Bolt hit the field with a load bang and dust blew everywhere. Few nights of beautyful polar cold air mass.

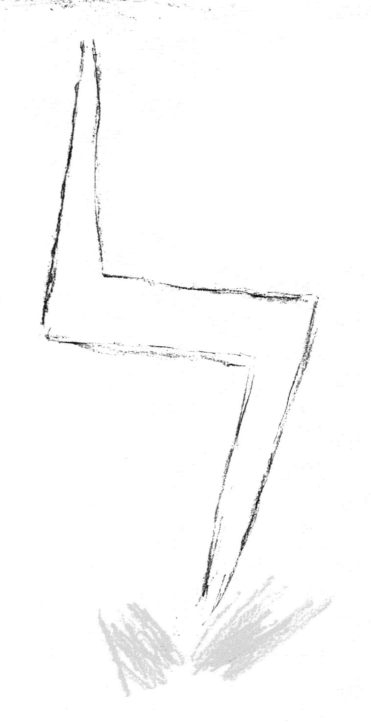

Myth of the Devil
on the Egg Farm

Where else would you find the Devil on the Egg Farm? In the Deviled eggs! on a March Saturday windy day in the old cooler barn on the egg farm. Sighted by two loaders of loadout, in differnce periods of time. Scientist say we live in the Illusion; science Illusion on the Egg Farm.

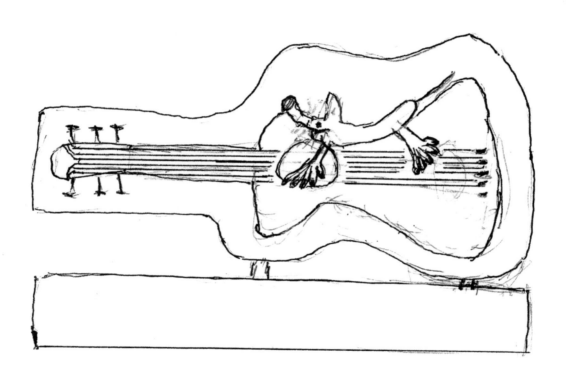

Study of a Mouse on
Bacon Chips & Water

Study of a mouse in a rented apartment. Feed the mouse bacon chips & water.

Slepte in a guitar case on top of a guitar. The mouse grew huge hands and huge feet and looked like a gorilla.

"Get out of Dodge"

"Floor Roller Rocket 88"

"Camper," mans best friend

"Mid-energy wonder nut"

"Camper, man best friend," will sleep with you when your sleeping on the floor with a sleeping bag. Camper squirrel sleeps under your sleeping bag and will lift the floor board up under the sleeping bag. Sometimes camper squirrel crys besides you at your sleep under the floor of your sleeping bag.

"Mid-Energy Wonder Nut," a working squirrel. Storage nuts during the winter in between the walls. The squirrel rolls the nuts down the staircase inside the wall. From the storage den to the eating den. At 3:00 A.M. you could hear the rolling nuts bounce down the staircase in between the wall.

"Floor Roller Rocket 88" will wake you up every morning at 8:00 A.M.. By rolling on the inside ceiling of the living room in the apartment. Below the T.V. on 24/7 soap operas nearby a couch to wake you up for work.

"Get Out of Dodge" squirrel has an high I.Q. for getting other squirrels for gathering. On the porch roof all the squirrels would stare at you. As you leave the apartment for work also all the squirrels would whistle at you, too!

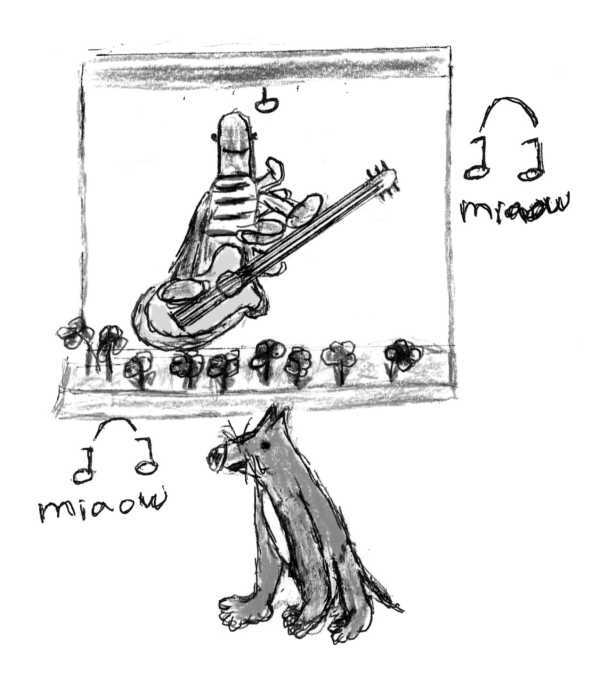

In a rented apartment playing on a guitar, cat blues to a stray cat. The stray cat miaowing along to the guitar note for note.

Recording a record of 'Crying Dog Blues'

Character
has been
changed for
protection
of the
innocent.

For 2 weeks at the same time every night. All the dogs in the neiborhood cry the crying dog blues. I can't get any sleep for 2 weeks. It's unreal!

Legends of Maine Gnomes Guardians of Treasurer of the Earth

In origin of Roman Literature of myths of Gnomes in accuracy. On a March windy night at the hunting camp on the lake, with the wind blowing into a tree hole making a musical tone. Fireplace lit cooking of chicken on a slab of wood. Sighted 3 Gnomes that were in likeness of a dream, in bright green and red clothes. Walking on air above the fireplace mantel toward the camp wall. Each Gnome in disappeance through the wall and left ashes into a pile.

MYTH OF A GNOME
RETURNED AS AN OTTER

The Working Otter would trot onto the porch of the hunting camp. The porch door of the hunting camp being unlocked, myth of A Gnome returned as an Otter. The Working OTTER would build his imaginary oar. From the memory of a 9 foot height, 23 foot length, 8 foot width ARK row-boat that was built in the 1930's. It would take the Working OTTER a whole day to build his imaginary oar. Just a plain old board 2 inches by 1 inch by 10 feet long. There was the pine board just standing there with the bottom portion of the board chewed off. When sighted, the board still standing in it's place on the hunting camp porch that the Working Otter had chewed, and the sawdust in it's place under the chewed portion. Being that of the nature of the Working OTTER in the spring, in the month of March 1975. Down by the lake where the hunting camp dwells.

Doll Monster in capture of the
Legends of Maine Gnomes

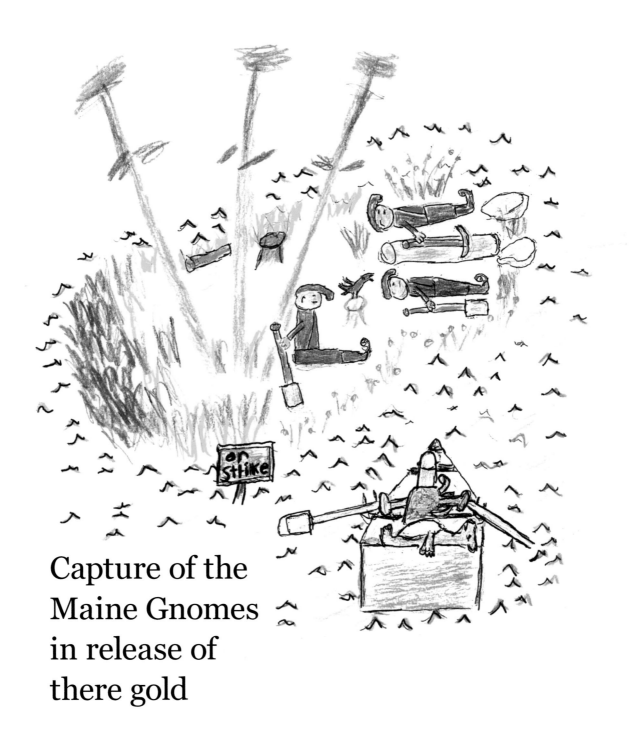

Capture of the
Maine Gnomes
in release of
there gold

Legend of Maine Gnomes, Broadcasted on one of the last 400 Nascar races summer 2022. (Announcer) The race drivers drive like 2 Gnomes with shovels, shoveling there way through and a Gnome with a pickaxe, picking his way through traffic.

At 4 years of age ran away to the circus. Left the diner where I was being babysitting. Crossed the highway and train tracks where the circus was. Sighted the elephant and stood there all morning watching his feeding and watering. At the circus I was sighted with the Red Yankee baseball jacket and caught and brought back to the diner.

Sprit of Thought to
Advance the Knowledge
of Flight

(CIA UFO Files)

Where at 4 yrs. of age every morning, use to entice a gray cat from the farm; on to the porch with milk and threw my gray boots at the cats head; where a song was written about this but the song was stolen;

Discovered a Blip 90 seconds in on the song '2120 South Michigan Avenue' sent in the cartoon & now history:

My Blue Berry Cow & My Blue Berry Milk

My Blue Berry Cow and My Blue Berry Milk in the morning. Wake up at 5:00 AM in the morning to milk my Blue Berry Cow. Get dressed in my farm clothes, go out to the barn to milk my Blue Berry Cow. Milking my Blue Berry Cow Squirt, Squirt, Blue Berry Milk in the milk pail. Milking my Blue Berry cow drip drop, drip drop, Blue Berry Milk in the milk pail. My Blue Berry Cow and My Blue Berry Milk in the morning.

Winning a trip to Disenyland from the Heartbeat of America Chevy Dealer.

Visiting Disenyland castle to pick up my stuffed doll 7 plate armadillo.

Where Disenyland called me at my apartment. We're not interested in the invention for Disenyland.

Pigeon bobbin his head up & down like a boxer
a top of the brick roof deco

6 jerseys green and black stripe worn in
a movie film. Made it on the film 'Your
In' on the Block for Bids.

Making a contest movie film of getting on a wave, on a round skimmer board of a wave coming a shore.

7 plate armadillo abducted by Elvis Tour Bus in Monmouth, Maine 1975. The armadillo made into a green neon light placed in a restaurant window on an Air Wolf show on cable T.V.. Ended up on the cutting room floor, before the Air Wolf Show came to market in DVD sale.

Fat, wild, gray, stubby, tailed squirrel that hopped. Lived in a forest in a branch pile nearby a pine tree stump. In back of a garage & 7 plate armadillo sign nearby a road.

Seabiscit

Visiting Seabiscits off spring, at the Woolworth Estates, Monmouth, Maine.

Fishing in front of Norman Lears rented camp on Lake Cobbosseecontee.

Nearby a dock of a New Jersey tenant girl in a bikini. Sunning herself under the summer sun on the dock. As a typical teenager fishing in a rowboat. Close by the dock watching the nature of the girl sunning herself on the dock.

(Scene, Merv Griffin Show Interview with Chris Everett) Merv Griffin asking Chris Everett: The name your wearing around your neck, Who is that?

Chris Everett answering Merv Griffin: It's nothing!!

(Narrator: Mailed, Bill Medly of the Righteous Brothers, Beverly Hills, CA. Lead Sheet Songs & Demos & Field Philosophy Theory at Lyra, return address to David. Where Chris Everett worn the return address 'David' around her neck.)

Famous cartoon sent to Katarina Witt

Famous quad skating spin at 11 years of age. Skated at a community pool ice rink under the lights. Skated around in circles for five minutes for the momentum of the spin. Then did the quad spin, it was like heaven.

Apprentice in restaurant cooking tomato sauce
from 5:00 A.M till noon; of a famous restaurant
where movie stars eat after their theater work
of acting;

Where there was a scuffle with the representative of the Sauage Co. at the diner. Later on the Sauage Co. representative came up to the house from the diner. Said the Boss said we had to buy his sauage. That night we had the sauage for supper.

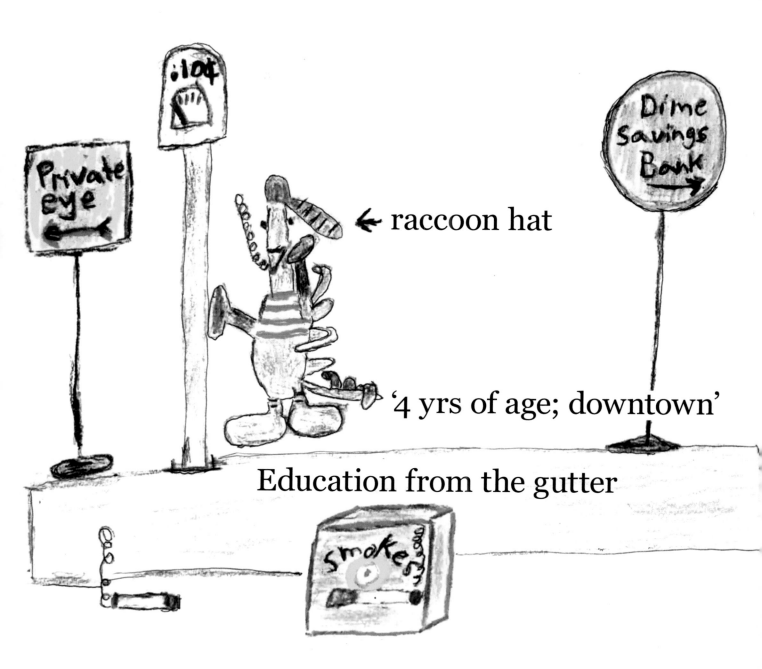

← raccoon hat

'4 yrs of age; downtown'

Education from the gutter

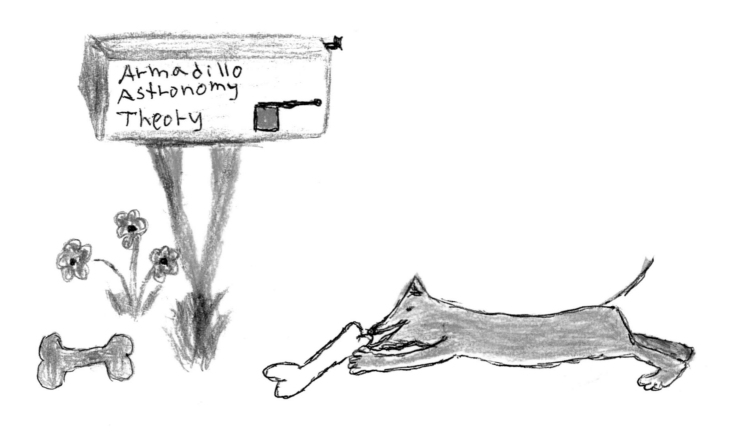

Dog Days of Summer

Neighbourhood dog enjoying a dirt flower driveway with his own bone. Instead of an abandon green rubber dog bone nearby the mail box.

Stray cat practicing scare tatics after observing another cat locked in an inclosed solor glass door extension.

Summer of 2023 an unreal 10" L by 1" dim. Bumblebee landed on my executive inclosed solor glass door extension. Later on that summer a smaller bumblebee with a red spot on the back of the

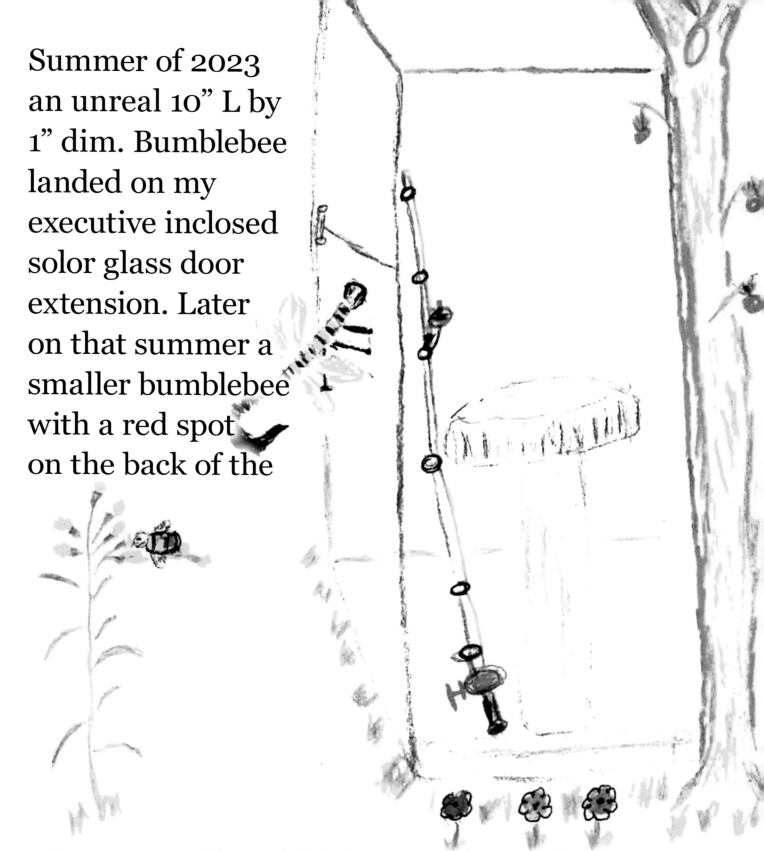

bee on a yellow field flower. Where I had 2 bumblebees nest. The bees scented the nests. Fish & game has the nest for study.

Young African American girl playing by hitting
the shines with a stick...

Foot on the table means mapped an Ice Planet like Mars at Lyra

80 foot shade tree

From the Radio, from the top archives October 2010. Young musicians.org Harpiest wrote the song "Doll Monster that ate the town" from the art contest book of 2000 of an armadillo & stair case drawing.

About the Author

Inventor, Cartooniest, Astronomy Theory of exploration of an Ice Planet at Lyra

Anthony L. Pace resides in West Gardiner, Maine

Printed in the United States
by Baker & Taylor Publisher Services